CLAY PIPES

by

EDWARD FLETCHER

Edited and Photographed by
Michael J. Coffin

© World Copyright 1977 - Southern Collectors Publications
7 Maple Road, Bitterne, Southampton.
I.S.B. No. 0-905438-09-4

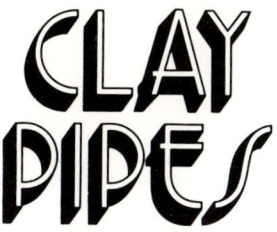

Clay Pipes

Edward Fletcher

CONTENTS

		Page No.
1.	Introduction	1
2.	History	2
3.	Late Victorian Pipes	4
4.	Fakes and Restorations	21
5.	Repairing Pipes	22
6.	Sites for Pipe Hunting	23
7.	Valuing Pipes	24
8.	Pipe Collections	15
9.	Pictorial Dating Guide	7

ACKNOWLEDGEMENTS

The Publishers would like to thank June Heath, of "Now & Then", Uxbridge and Andy Payne of "The Bottle Shop" Southampton for their kind assistance in the preparation of this Book.

1 Introduction

The primary aim of this little book is to encourage greater interest in collecting clay tobacco pipes by providing in a single volume all the information needed to make a start on this fascinating hobby. Although it is true that many experienced bottle diggers and collectors tend to include clay tobacco pipes in that miscellaneous bag of odds-and-ends known as "other dump finds", it is equally true that clay pipes are already of enormous interest to beginners at Victorian relic hunting. The two main reasons for this are: Firstly, that even the most inexperienced beginner who digs on the most overworked site will quickly find his or her first pipe bowl. Secondly, that despite the ever-rising prices paid for all manner of bottles and relics these days, clay tobacco pipes remain among the least expensive items of *genuine* Victoriana to be found on dealers' shelves.

These two considerations certainly make clay pipes an attractive proposition for the newcomer, but it is two rather less widely appreciated facts about pipes that arouse the enthusiasm of the handful of dedicated professionals who make their livings by dump digging. Firstly, the presence of decorated tobacco pipe bowls in any late Victorian dump (or the presence of decorated stems on the surface of an undug dump) is a positive indication that the dump holds rare bottles. Secondly, there are a small number of extremely rare clay tobacco pipes to be found in British dumps which command prices equal to those paid for top bottles and for which the true professional digger is prepared to hunt long and diligently. Thus, at both ends of the scale - absolute beginners and hardened professionals - there is already much enthusiasm for these relics of smoking history. If in the chapters that follow I can encourage others, perhaps middle-of-the-road collectors who have found large numbers of pipes in the past, to take a greater interest in such finds I shall consider the primary aim of this book to have been successfully achieved.

A word on sites: So far as the average digger-collector is concerned clay tobacco pipes fall into two categories:- those which are found in turn-of-the-century bottle dumps (1880 - 1920), and those which are found by archaeologists during excavations on post-medieval sites. Mr. Average Digger-Collector (if such an animal exists) almost invariably limits his interest to the former. Similarly, many archaeologists seem under the impression that the only clay tobacco pipes worthy of scholarly attention are those made in the 16th, 17th, 18th, and very early 19th centuries. How wrong both groups are! There are thousands of non-archaeological sites where non-scholarly collectors equipped with a few simple tools can find and add to their collections many examples of early pipes which, although lacking the decorative and ornamented shapes of late Victorian specimens, do nevertheless complement the type of collection likely to be owned by a dump digger. Again, archaeologists could, I feel quite certain, dramatically change their opinions on clay tobacco pipe history if persuaded to turn their learned attentions to "Viccy tips". Authoritative and scholarly work on the clay tobacco pipes of the late 19th and early 20th centuries is long overdue and - if popular interest is any guide - urgently needed. Perhaps the photographs of "amateur finds" which accompany the text of this book will convince those archaeologists who happen to turn the pages that such a project is worthy of their time and effort.

2 History

It all began in the late 16th century when European voyagers to the New World picked up, and brought back with them to the Old World, the American Indian habit of burning and inhaling the smoke of the "uppowoc" plant, a word which the Spaniards translated as "tobacco". One of many accounts describing how the Indians used tobacco was written in the 1580's. It shows clearly that the idea of "the pipe" was also borrowed from the New World:

> "When they (the Florida Indians) travel they have a kind of herb dried, which with a cane and earthen cup in the end, with fire, and the dried herb put together, do suck through the cane the smoke thereof". (John Sparke 1589).

Historians differ on who is to be credited with introducing tobacco to England. Some say Drake brought it back from the West Indies in 1573; others that Raleigh first returned with it after his voyage to North Carolina in 1585. Captain Richard Greville and John Hawkins have also been suggested as candidates. Whoever the "brave sea captain", there is no argument that by 1598 tobacco smoking was a well-established habit throughout southern England. A German writer reported in that year that - "the English are constantly smoking Tobacco. They have pipes on purpose made of clay. They draw the smoke into their mouths which they puff out again through their nostrils like funnels along with plenty of phlegm and defluxion from the head".

The habit, which was probably both social *and* medicinal, proved equally popular with the fair sex. Writing of his visit to England in 1618 the Venetian Ambassador reported:

> "One of the most notable things I see in this Kingdom is the use of the Queen's Weed, properly called tobacco, whose dried leaves come from the Indies. It is cut and pounded and subsequently placed in a hollow instrument, a span long, called a pipe. Women as well as men smoke night and day".

Seamen who took up smoking during their voyages invariably settled - when their wandering days were over - in coastal towns, and it was the concentration of large numbers of smokers in and around these ports (coupled with the fact that imported tobacco arrived there) that led to the establishment of pipe-making industries in the same places. London, Bristol, and most of the large ports on the south and west coasts were among the first to produce mould-made clay tobacco pipes. As the habit spread northwards it was in ports such as Harwich, Hull, and York (a major port in those days) that pipemakers set up the tools of their trade.

By 1619 these men had become an organised body with a Royal Charter of Incorporation. The most powerful contingent was the Company of Pipemakers of Westminster which had rights of control throughout England and Wales, though in practice the London makers were only able to enforce their monopoly within a twenty mile radius of the capital. The numbers employed in the trade by the mid-17th century can be gauged from the wording of a petition presented to Parliament in the 1640's by pipemakers complaining about the introduction of a tax on clay tobacco pipes:

> "Before the late Act of Parliament for laying a duty on tobacco pipes near 1,000 poor people in London and Westminster lived by tobacco pipe making, who for want of such employment are become beggars. Several thousands of other tobacco pipe makers throughout England and Wales are in like manner ready to starve

for want of employment, not more than a fifth part of the pipes made before the imposing of the said duty are now being made, foul pipes being oftener burnt than formerly, and the sea trade almost lost by reason of the said imposition".

Other major considerations governing the establishment of a pipe-making industry in a particular town were the availability of suitable clays and an adequate supply of fuel for firing kilns. The best pipemaking clays were (in the 16th century) found in Dorset and Devon, and the owners of these deposits supplied clay not only to local makers but also for makers in London where a shortage of suitable clays was a constant problem. London was equally short of fuel, especially in the 17th century when coal-firing became common, but despite these two disadvantages the London makers flourished thanks to their monopoly and the vast market provided by the ever-increasing population of the capital. In other parts of the country a combination of good local clays and supplies of cheap coal ensured a thriving industry. By 1750 these major centres of manufacture were: London, Plymouth, Bristol, Broseley, Chester, Hull, York, Carrickfergus (N.Ireland), and Edinburgh. There were, of course, thousands of makers in other areas, (*Every* town had at least one maker in the 17th and 18th centuries) but there is no doubt the *major* centres influenced the shapes and styles of pipes made in the less important regions. Thus, by comparing pipes found during searches and digs with pipes of *known date* from these major centres it is usually possible to arrive at an *approximate* date of manufacture for almost any British-made pipe dating from 1590 to 1850. Readers who wish to establish an approximate date of manufacture for a pipe without doing further research should consult Figs 1 - 10 which show the evolution of bowl shapes. Readers who wish to establish a more accurate date should read *local* archaeological reports which often include lists of makers and more detailed information on local bowl shapes.

Markings and decoration: When compared with the superbly ornamented late 19th early 20th century pipes found by dump diggers, these pre-1850's pipes at first appear rather drab, but *some* were quite elaborately decorated in the 18th century. Notable are the Carrickfergus specimens which carried regimental emblems as early as 1750, and those from other major centres bearing town and city coats-of-arms. Bowl markings on pre-1750's specimens are usually limited to the initials of the maker (which should be read left-to-right with the pipe held in the normal smoking position) or at best the full name of the maker impressed along the stem. A few are found with markings (usually initials) on the rear or side of the bowl, while quite large numbers of very early specimens have "rouletted" marks around the top of the bowl. Any pipe which can be positively dated before 1800 and which has decorations other than those mentioned here should be considered extremely rare. Pipes bearing dates are equally rare.

Moulds and methods of manufacture: The best way to study methods of manufacture is to visit museums where moulds and other pipemaking tools are preserved. (See Part 8). Briefly, the prepared clay was formed in a two-piece iron or brass mould. The hole in the stem was achieved by pushing a piece of wire into the closed mould, and the inside shape of the bowl was produced by pushing a plunger into the top of the mould. (This was carried out with a "screw" in a few cases, and in many cases on Dutch pipes). Decorations such as "rouletting", makers' initials, and final finishing were carried out with a variety of wheels and knives after the partly-formed pipe had been removed from the mould prior to firing. The unfired pipes were then carefully placed - hundreds at a time - into the kiln where they were fired for several hours to produce the finished wares.

Exports and imports: Most of the very early clay tobacco pipes found on colonial sites in America, Canada, and the West Indies were made in London. By 1700 the importance of Bristol as an international manufacturing centre began to grow and many pipes made in Bristol between 1700 and 1750 have been found overseas. Liverpool's pipemakers took a substantial slice of the American market in the early 19th century, as did makers in Glasgow. Pipes from many British towns have also been found by diggers of Victorian dumps in Australia, South Africa, and New Zealand.

So far as imports are concerned only the Dutch presented competition to British makers before 1800. Dutch pipes (see Fig 10 for bowl shapes) have been found in large numbers at various sites in London, Harwich, Plymouth, Hull, and other ports. In the 19th century French pipes - many of them richly decorated - began to find a growing market on this side of the Channel. Best known to dump diggers are the figural pipes made by Gambier which turn up regularly in late Victorian dumps. Red specimens made by Fiolet of St. Omer are also found, as are the wares of several other French makers. Most are richly decorated.

3 Late Victorian Pipes

One hears all sorts of figures quoted within the dump digging fraternity as "official" estimates of the number of *different* styles of pipe bowl so far recovered from 1880's - 1920's sites. 2,000; 5,000; and 10,000 seem to be the figures most widely bandied about, and I invite readers to form their own opinions as to which is closest to the truth. (Please note that I am talking about *different* types; actual numbers of pipes must run to tens of thousands). I *do* know that in 1972 professional digger John Webb of Essex had a personal collection of almost 1,000 different specimens - all decorated or figural bowls - and all, alas, stolen (and never recovered) during house-moving operations. I also know that John's digging activities were at that time confined to London, Essex, and Kent so the figure "2,000" as a *national* total is very far short of the mark. Numerous dumps in all parts of Britain have been excavated since 1972 and although most of the army of diggers now following the hobby will never achieve John Webb's standard of professionalism, their combined efforts must surely have produced as many different types. What prevents me sticking my neck out and "guesstimating" that 5,000 is the closest-to-true figure are the *thousands* of sites still awaiting the bite of a digging fork, (See my earlier book: "Where to Dig Up Antiques") and the thought that twelve months from now even the figure "10,000" could look very conservative . . . . . .

How can this apparent explosion of ornate pipe making in the late Victorian period be accounted for? Well, in the first place, I don't think there was a sudden burst of activity by pipe makers in the 1880's. Our views as dump diggers are distorted by the fact that we have very few pre-1880's dumps to explore. If we had I'm sure we would find a *gradual* build-up in the number of ornately decorated pipes from the very beginning of Queen Victoria's reign. Factors contributing to the build up were:

1. An ever-increasing population and a growing popular taste for tobacco.
2. A steady improvement throughout the 19th century in transport facilities which brought wider markets within the range of pipe makers. In the 18th century distribution was limited to the range of a fully-loaded packhorse; in the late 19th century a network of railways provided cheap transport which enabled provincial makers

3. to have their products on sale in London or any other major city within twenty four hours. Improvements in world-wide shipping facilities also provided larger markets overseas.
3. Competition - as a direct result of improvements in transport - forced pipe makers to produce more elaborate and eye-catching designs.
4. The Victorian era was the *golden age* of ornate decoration. Everything from the town-hall facade to the humble clay tobacco pipe received the attentions of artists, designers, and craftsmen who were apparently incapable of leaving the surfaces of any object plain and simple. When the "plain and simple" came back into fashion in the 1920's, clay tobacco pipes reverted to their earlier austere designs.
5. Advertising - a force all too familiar to us today - was just beginning to exert its insidious pressure when Queen Victoria came to the throne. The Great Exhibition of 1851 showed the world that good advertising could improve trade and the idea caught on very quickly. Apart from newspapers and hoarding what better, cheaper method was there of telling the world about your public house, your forthcoming event, your latest music-hall success than by utilizing the surfaces of the ubiquitous, inexpensive, and ready-to-hand clay tobacco pipe? If cigarettes had not largely replaced pipes I'm sure present-day ad men would be writing their copy and their jingles on the sides of smoking briars instead of on the sides of football pitches and underground escalators.

By 1880 the practice of including makers' initials on pipe bowls had steeply declined while the area of distribution available to each maker had greatly increased. This encouraged *some* makers to include their full names *and* the name of their town on the stems of their pipes, but many more abandoned such marks of identification altogether. For this reason it is extremely difficult to attribute late Victorian pipes to individual makers. Collectors must resign themselves to attempts at dating their finds by other methods, and the best is undoubtedly to establish the date of the dump from which a pipe was dug by dating bottles and other items found alongside the pipe. Another method which works with pipes of certain shapes and designs is to consider why the pipe was made that way. A bowl in the shape of the "Rocket" Locomotive, for example, must surely have been made to commemmorate the arrival of steam passenger trains. Similarly, a bowl bearing portraits of Generals Bullers and Kruger was almost certainly made at the time of the Boer War, while those depicting the features of politicians and music-hall stars are most unlikely to have been made many years *after* these personalities fell from public favour.

This method of "dating-by-event" can give an approximate year of manufacture for many late Victorian figural pipes, but certain figurals cannot be dated in this way because they were produced over long periods of time. The most obvious examples in this category are the figurals of Queen Victoria and other members of the Royal Family. Their popularity caused pipe makers throughout Britain to have moulds made in their images for many years, especially during the period between the Queen's marriage to Prince Albert and her death. Another figural shape which had a long run of popularity was the "Negro Boy", the traditional trade mark of the tobacconist for more than a century. It was used by the majority of pipe makers who produced figurals and examples have been found by diggers in most parts of Britain. Father Christmas figurals - some of which were made with enormous bowls - and those portraying ever-popular characters such as Dick Whittington, Ally Sloper, characters from Charles Dickens' novels, and

KEY TO ILLUSTRATIONS

1.)
2.) Posies of Flowers
3. Plain Bowl
4. Naughty Nineties — Girl on Chamber Pot
5. Basket Weave
6. Two Dice
7. Decorative Bowl — Use of Screw Thread Unknown
8. Acorn
9.)
10.) Hand Holding Bowl
11. Skull
12. Skull (by Gambier of France)
13. Princess Alexandra
14. Hussar
15. Negro Head (by Charles Crop)
16. "Buffalo Bill" Cody
17. Ally Sloper
18. Variation on Hussar
19. Queen Victoria
20. Bacchus
21. Cock
22. Eagle Claw
23. Man Relieving himself against Tree Trunk.
24.)
25.) Complete Pipes with Slogans on Stems.
26. Early style Bowl — CIRCA 1620
27. Figures on Bowl — often show Footballers, Cricketers, Boxers etc.
28. Hoof

Fig. 1

British bowl shapes 1580 - 1850.
Simplified. Compare with regional variations shown in Figs. 2 - 9.

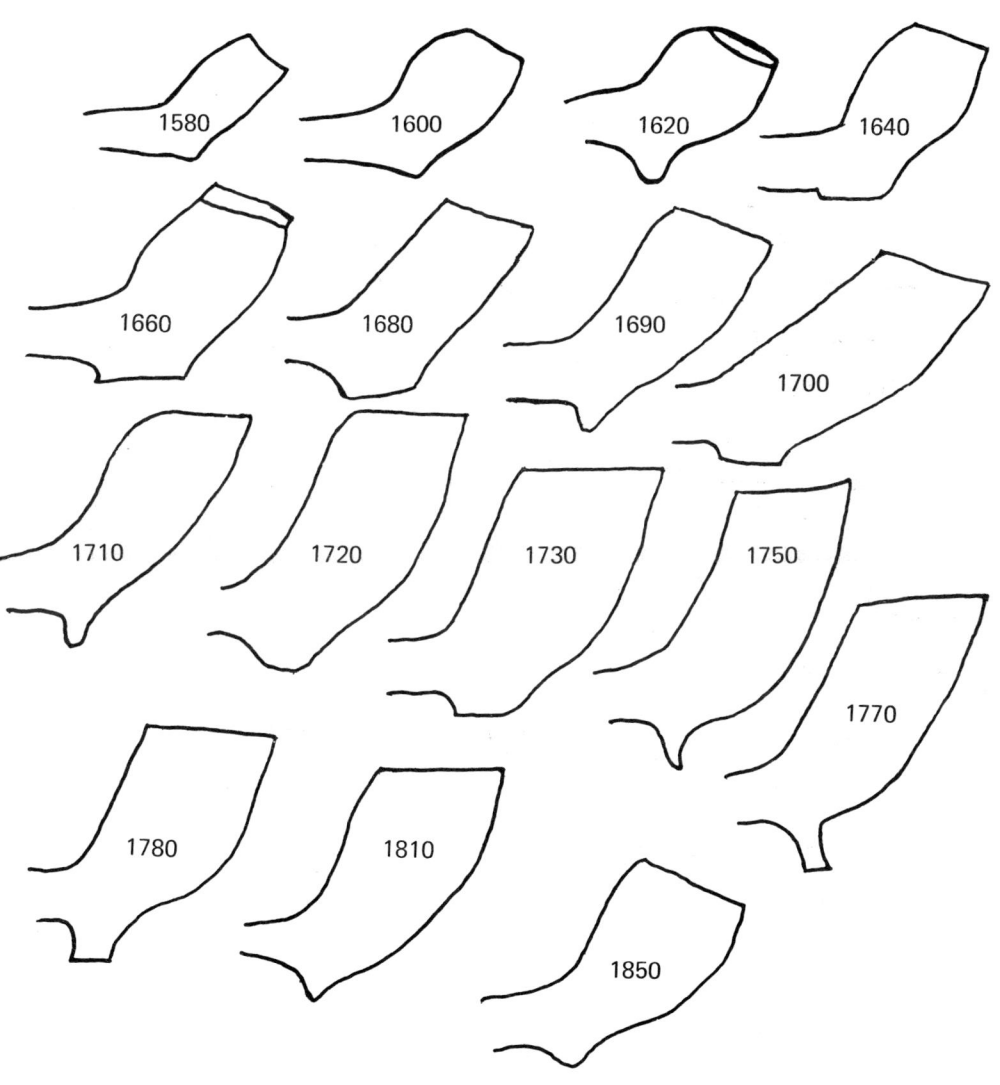

Fig. 2

London Types

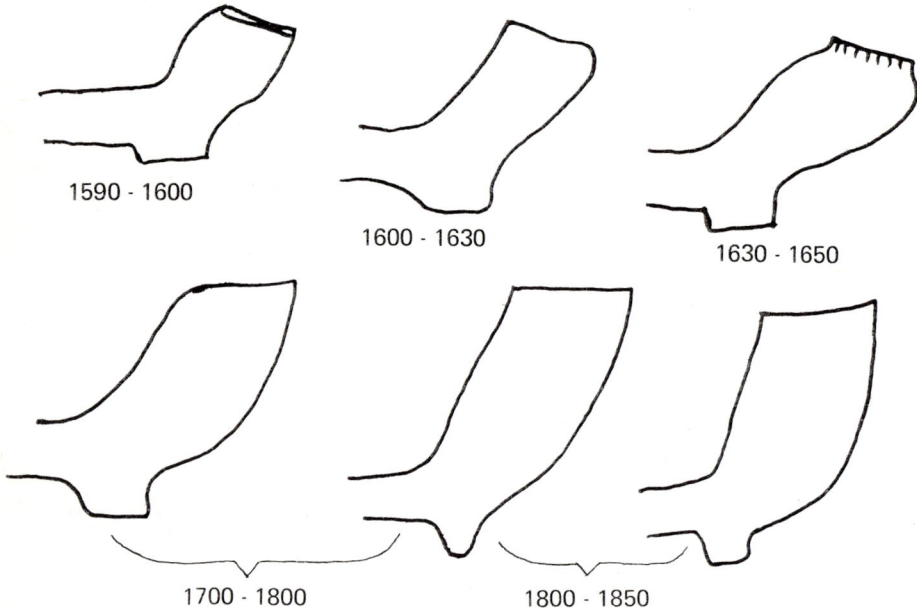

1590 - 1600

1600 - 1630

1630 - 1650

1700 - 1800

1800 - 1850

Notes:- Polished finish on bowls fairly common after 1700. Makers marks on almost all specimens 1600 - 1700.

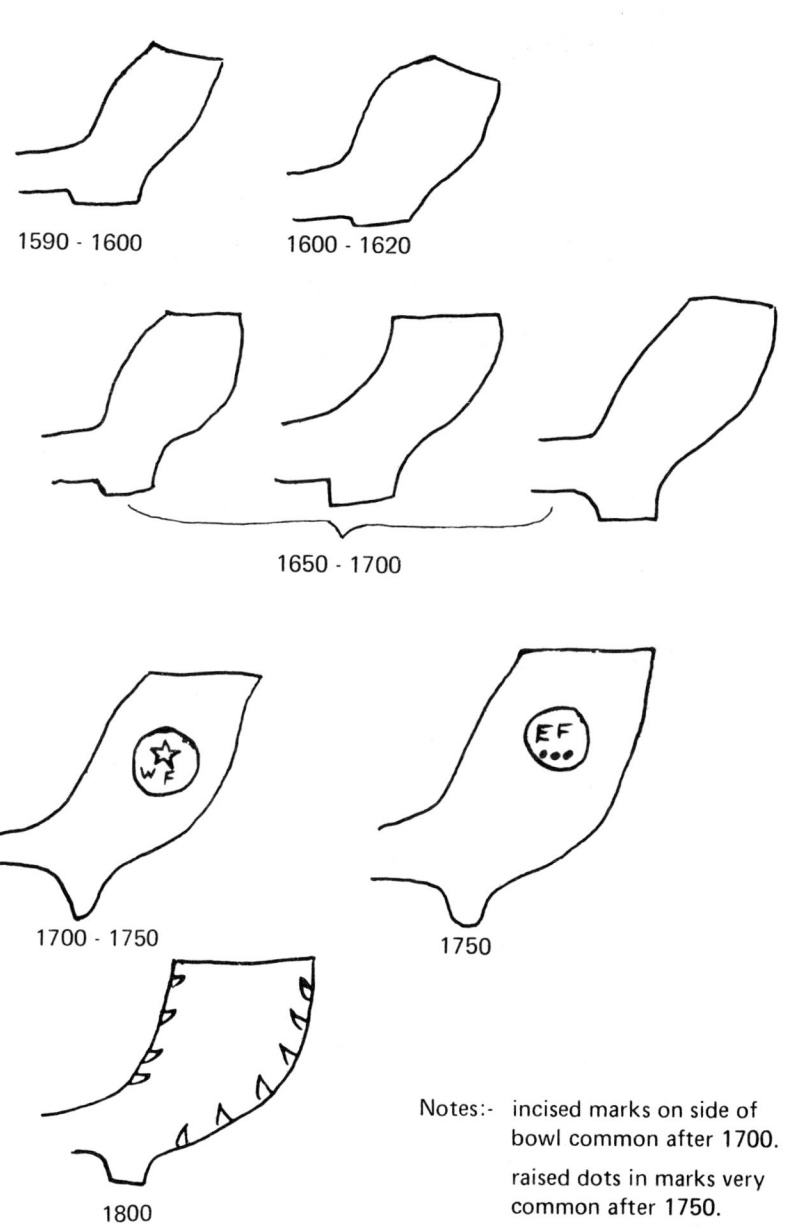

Fig. 3
Bristol Types

1590 - 1600

1600 - 1620

1650 - 1700

1700 - 1750

1750

1800

Notes:- incised marks on side of bowl common after 1700.

raised dots in marks very common after 1750.

Fig. 4
Plymouth Types

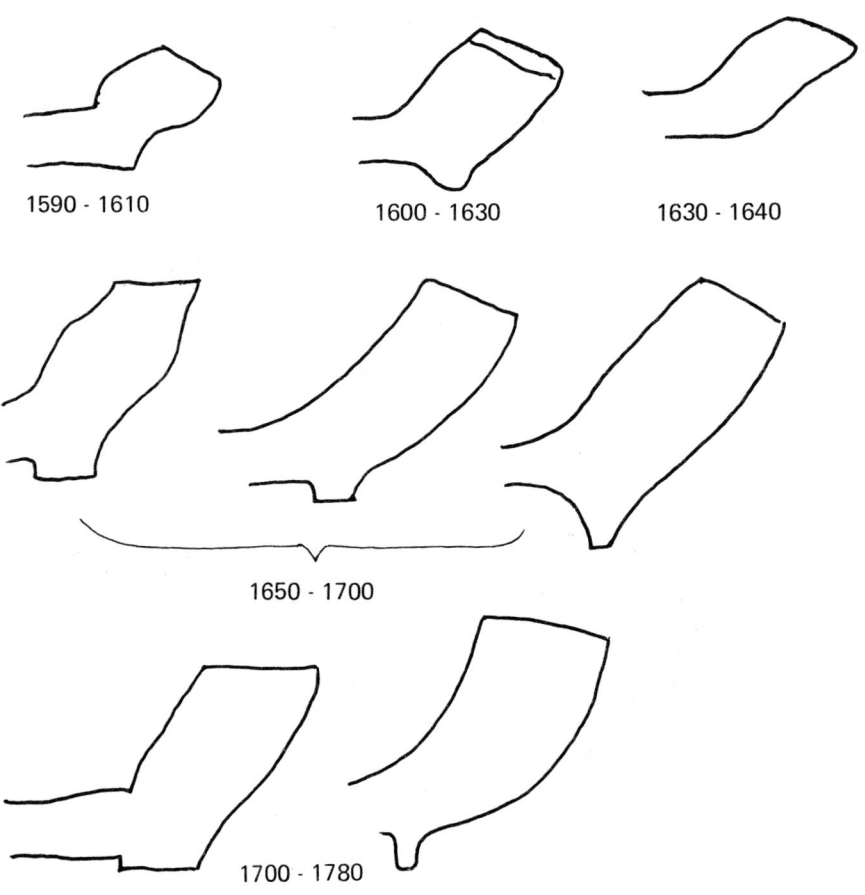

Note:- Forward projecting bowls common until 1700.

Photo Page 1

Photo Page 2

12

Photo Page 3

21

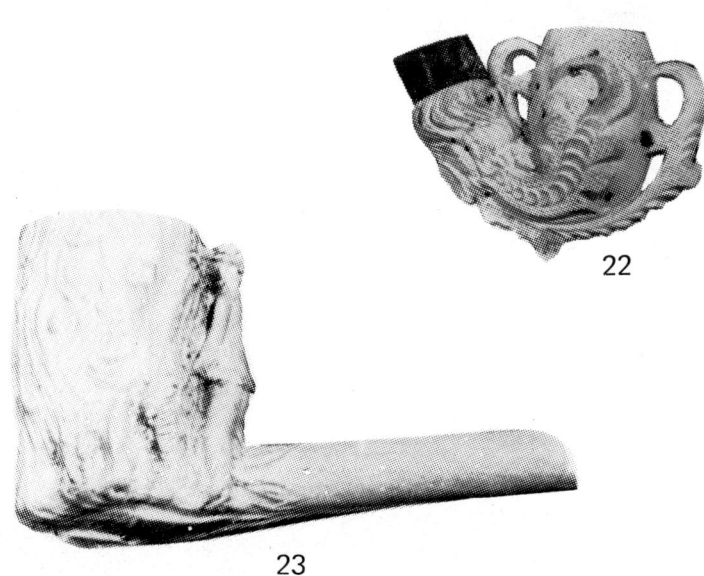

22

23

Photo Page 4

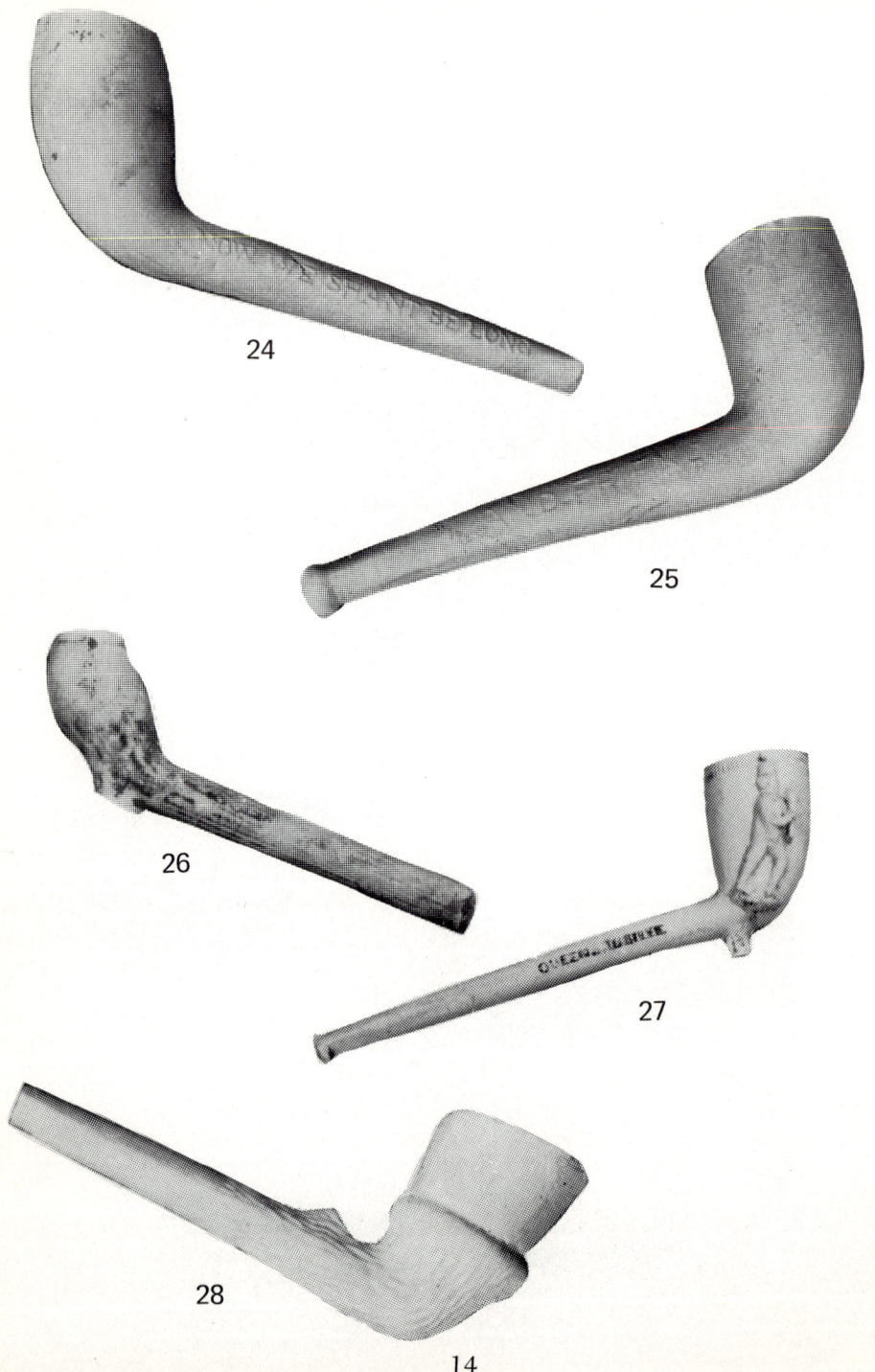

Fig. 5
Broseley Types

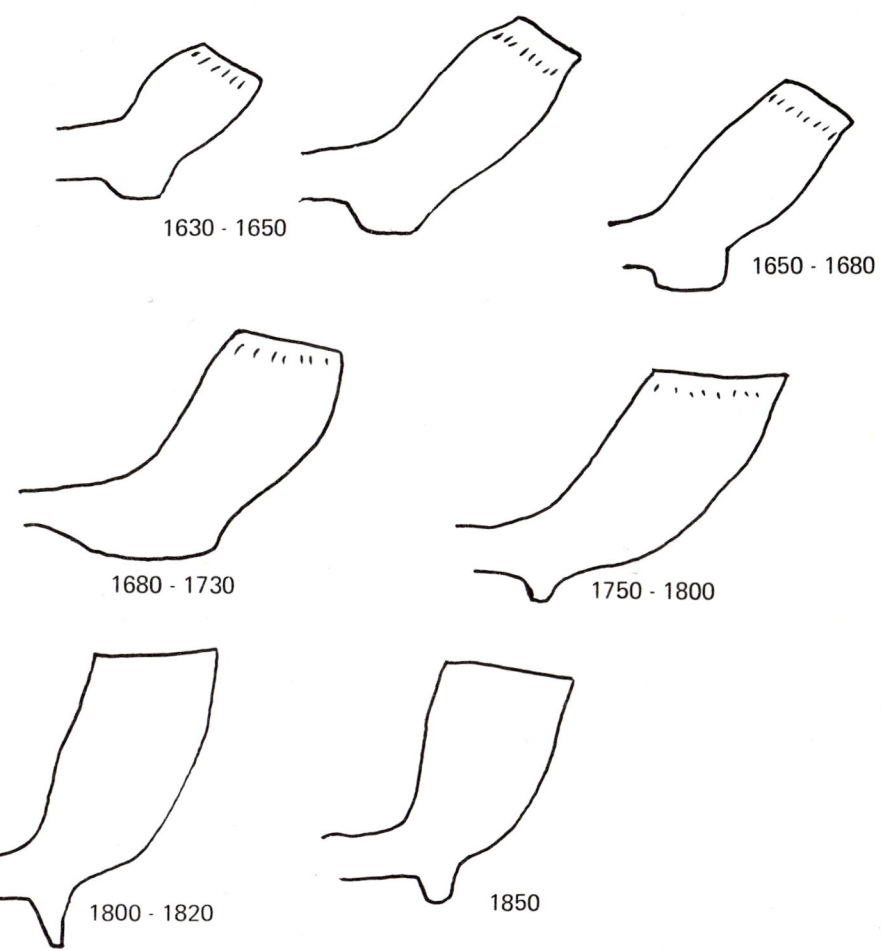

Notes:- large flat spurs on bases of bowls very common up to 1730.

— most Broseley pipes have some form of decoration around top of bowl up to 1800.

Fig. 6
Chester Types

1600 - 1650   1650 - 1700

Note:- Many Chester pipes have markings in relief (embossed)

Fig. 7
Carrickfergus Types

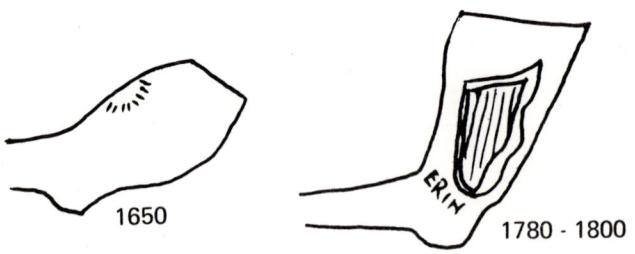

1650   1780 - 1800

Notes:- Bowl marking fairly common all periods.
Patriotic markings very common after 1750.

Fig. 8
Scottish Types

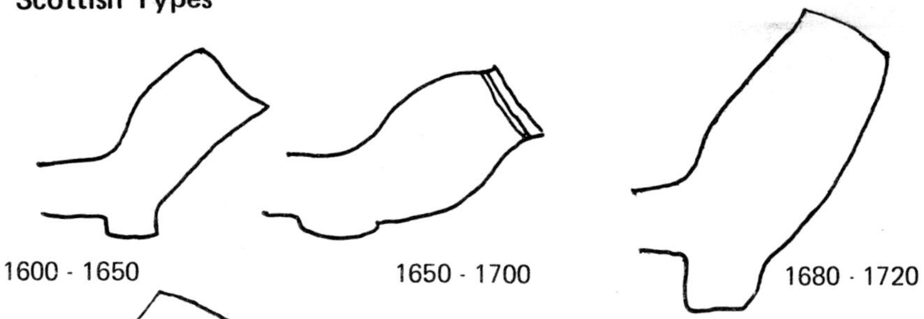

1600 - 1650   1650 - 1700   1680 - 1720

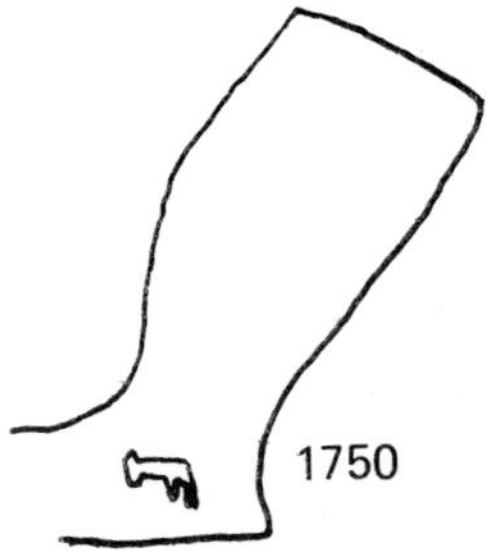

1750

Notes:- Many Scottish pipes have a yellowish colouring.

Polished specimens are rare.

Makers initials are after in relief (embossed).

Fig. 9
Yorkshire Types (York and Hull)

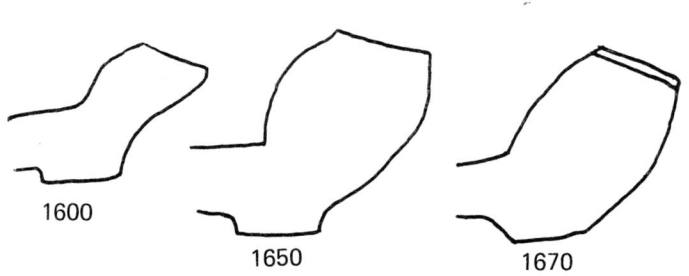

1600   1650   1670

Note:- Bulbous bowls similar to that shown for 1650 are fairly common in Yorkshire.

Fig. 10
Early Dutch Types found in Britain

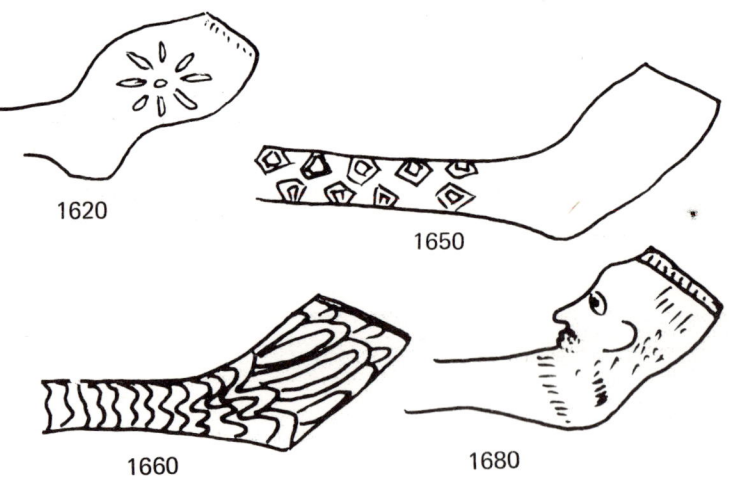

Notes:- Decoration around the top of the bowl is <u>very</u> common on Dutch pipes.

Clay colour is usually light grey or yellow.

Bowls are usually highly polished.

Bowl and stem decorations are very common.

biblical or mythological characters were also made over many years. The French maker Gambier was famous for his "Jacob" pipes and these were copied by some English makers over long periods.

As mentioned in the previous chapter, smoking was considered an acceptable feminine habit in the 16th and 17th centuries. By the late 19th century the standards of society had changed. No lady who valued her good name would have smoked in public, and it was not until the "Gay Twenties" that large numbers of women adopted the habit once again. In Victorian households gentlemen retired to the smoking room or waited until the ladies had "withdrawn" before they lit up their pipes. This practice led to the manufacture and sale of quite large numbers of clay tobacco pipes which, in those days, must have been regarded at best as "for men's eyes only"; at worst as "obscene and pornographic". They included specimens in the shape of chamber pots, young girls sitting on lavatory pedestals, nudes, phallic representations, even some depicting acts of copulation. In private houses they must have been carefully hidden from the eyes of wives, daughters, and maidservants, which probably accounts for the fact that more have survived intact and above ground than have been found in refuse dumps. The odd specimens that do turn up in dumps command high prices; I know one digger who was paid £50 for a copulation specimen with a repaired stem - and he was innundated with requests for more of the same for many months afterwards.

Another group of pipes which clearly reveal the masculinity of the habit in the late 19th century are the masonic specimens, commonest of which is the R.A.O.B. pipe with buffalo horns. Many hundreds of different models of this pipe have been recovered, an indication of the importance of the Royal Antediluvian Order of the Buffaloes in those days. Members of the Ancient Order of the Druids used pipes in the shape of acorns, while figural pipes of Robin Hood and those with bowls in the shape of trees were probably favoured by the Ancient Order of Foresters.

While large numbers of pipes were undoubtedly sold by tobacconists' shops, even more were on sale in public houses. Visitors to York's Castle Museum should look out for the period setting which shows a typical turn-of-the-century public house. On one of the walls is a display card filled with pipes. These were apparently sold at "one halfpenny each", but in some public houses they were given away free of charge to regular customers. Specimens in this category usually included some form of advertising. A pub called "The Bunch of Grapes" would have plentiful supplies of pipes with bowls in the shape of a bunch of grapes, if only to remind "hungover" customers on "the-morning-after-the-night-before" where they had spent their hard-earned cash!

The reproduction of older pipe styles also enjoyed some popularity in the late Victorian era. Substantial numbers of plain pipes of typical 1780's shape have been found in 1880's dumps around London - far too many for them to have been very old pipes that had taken a century to reach the dustbin. I'm sure there must have been a decreasing but persistent demand for these old styles in spite of the attractiveness of the latest offerings in tobacconists' shop windows. Old men who recalled the styles of pipes used by their fathers and grandfathers were no doubt capable of resisting those "newfangled ideas". A parallel can be seen today in the large numbers of cigarette smokers who insist on buying brands that have been on the market since before the Second World War. (I know of no way to distinguish between 1790's specimens and 1890's reproductions other than by confirming the sites on which the pipes were found).

Miniature pipes - some copies of earlier styles, other copies of the latest figurals - seem, on digging evidence, to have been popular around the turn-of-the-century. It is interesting to note that the *internal* diameters of bowls on these specimens are all very similar and that they approximate the diameter of a cigarette. Perhaps pipe smokers who changed to the new cigarettes that were coming onto the market at that time did so via a pipe that was just sufficiently large to hold a cigarette in its bowl. If that was the case these miniature pipes ought to be regarded as the forerunners of cigarette holders.

It would be of enormous benefit to pipe collectors everywhere if those members of bottle collectors clubs who have specialized in late Victorian clay tobacco pipes formed themselves into a separate group and pooled their information on types and numbers of pipes recovered from dumps and in their collections. I'm sure that pipe collecting, as a "sub-hobby" within the bottle collecting fraternity would attract even greater numbers of devotees if one brave soul would organise a "pipes only" swap meeting or exhibition which would surely lead to the inauguration of such a club. One of the first tasks it might undertake would be the classification of late Victorian pipes into different categories, just as bottles are grouped as mineral waters, beers, inks, poisons, etc. Such classification would encourage more pipe lovers to specialise in particular categories, would lead to a greater wealth of detailed knowledge and, in the long term, would put pipe collecting on an *equal* footing with bottle collecting. As a first tentative step on that road I offer the following classification as a *starting point.* I stress that it *is* merely a starting point - from which I hope all readers will draw up more extensive classifications. If these are pooled whenever individual collectors meet it will not take long to establish a comprehensive pipe index.

1. Abstract designs - pattern on bowls.
2. Basketwork designs.
3. Floral and botanical - including leaves, wheatsheafs, trees, tulips, fruit, etc.
4. Heraldic - including town and city coats-of-arms, flags, national emblems, etc.
5. Political and royalty - including heads of prime ministers, Queen Victoria, etc.
6. Masonic - R.A.O.B. etc.
7. Sporting events and equipment - including bowls with footballs, etc. beneath.
8. Women - including music hall stars, etc.
9. Clothing - including hats, boots.
10. Transport - ships, railways, etc.
11. Claws and hands - supporting bowls.
12. Military - including generals, regimental badges, etc.
13. Pornographic.
14. Skulls.
15. Advertising - including negro heads, public house signs.
16. Commemoratives - including events, exhibitions, etc.
17. Animals.
18. Mythological, biblical, and fictional characters - Bacchus, Jacob, Dick Whittington, etc.
19. Miniatures.
20. Extra large specimens.
21. Stub-stemmed pipes designed to take mouthpieces made from wood, amber, etc.
22. Specimens bearing dates, place names, addresses, slogans, and other wording.

4 Fakes and Restorations

Readers who also happen to be smokers will be aware that there are quite large numbers of "reproduction" clay tobacco pipes on the market. They are made by firms supplying tobacconist's shop sundries and are primarily intended for use by pipe smokers who, I am told, find they get a cooler smoke from a clay pipe than they do from a briar. It is hardly surprising that these reproductions have proved popular; most of them are made in *original* Victorian moulds, a number of which have survived outside museums, and the makers have selected several eye-catching specimens including Queen Victoria's head, the Negro Boy, General Gordon, and other well-known figurals which even today are still sufficiently striking to catch the attention of the average pipe smoker whether or not he collects old pipes.

Unfortunately, they have also attracted the eyes of many newcomers to Victoriana collecting who have bought thinking them to be *genuine* pre-1900 items. Certain unscrupulous dealers have even gone to the trouble of "ageing" these modern items and passing them off as "dump finds" at inflated prices. Quite experienced collectors have been caught by the fakers who, unless their activities are stopped, could ruin the hobby by flooding the market with these spurious wares. For this reason I urge all readers to be "on guard" when buying clay tobacco pipes and to apply the following rules before parting with money:

1. If possible buy from a reputable dealer or digger within the bottle collecting fraternity who values your future custom and his own good name.
2. If in doubt about any specimen - perhaps because the price seems low or the condition of the pipe seems too good to be true - examine it carefully. If the mouth end of the stem has been scraped, suspect a "faking job". The stems of modern reproductions are usually dipped in a red, waxy substance which prevents the dry clay sticking to the smoker's lips. In Victorian times this inherent problem of clay tobacco pipes was overcome by dipping the stem into a mug of beer before lighting the pipe. The ends of *genuine* pre-1900 pipes were never waxed.

Look carefully into the bottom of the bowl and also into the hole in the stem - both places where a faker has difficulty in reproducing the effects of age and use. The usual method of faking age and long use is to rub the entire surface of the pipe with a mixture of cigarette ash and instant coffee moistened with a little water. The treatment can certainly leave a pipe looking surprisingly old - but the effect is very easy to wash off. Moisten the stem and wipe it with a paper tissue; if the "dirt and grime" comes off very easily and leaves a clean white surface beneath, it is most unlikely that the "muck" has been ingrained by nearly a century of burial. Suspect any pipe which has a very clean hole in the stem or one with a bottom to its bowl that does not look as if it has been well burned. Fakers find it very difficult to work ashes and coffee into these small spaces.

Many diggers go to great lengths to clean their finds before offering them for sale. Treatments include long soakings in "Domestos" or similar chlorine-type household cleaners. The smell lingers long after the pipe has dried out and it is therefore a good idea to "sniff" a pipe before buying if it looks too clean. In my opinion it is an even better idea to leave *some* of the "dump dirt" on all genuine "dug" pipes as this proves their authenticity.

Another way in which many fakes can be spotted is by comparing them with genuine pipes originally made in similar moulds. There is no doubt that modern reproductions lack the fine detail found on Victorian pipes even when the moderns have been formed in pre-1900 moulds. I believe this has something to do with the method of firing. Modern specimens are fired in electrically-heated kilns which seem to shrink the clay rather more than the earlier coal-fired kilns did. This results in a poorer reproduction of the moulded shape with loss of fine detail especially noticeable on facial features and hair in figural pipes.

Of course, the ultimate protection against the sale of reproductions as "genuine Victoriana" would be a method of *indelibly* marking all reproductions. At present such marks (e.g. the stamps found on the bases of reproduction ginger beers) are far too easy to remove. Almost as good would be a *complete* list of all reproductions on the market. If such a list was available to collectors they would be "on guard" whenever confronted with a pipe on the "suspect list". The compiling of such a list ought to be one of the first tasks attempted by a pipe collectors' club.

5 Repairing Pipes

Repaired pipes are far more acceptable to collectors than are repaired bottles, the reason for this being that more than 90% of pipes recovered from dumps have broken stems. (The stems were broken *before* the pipes were thrown away so check when buying a broken-stemmed specimen that there is ingrained dirt on the break). It is not absolutely essential to repair broken stems; if you display your collection on black pegboard it is in fact easier to fit the pipes into the holes when the stems are broken. However, there is no doubt one's display is much improved by having a number of complete specimens on show alongside the pegboard.

Before attempting any repair it is essential to clean around the damaged area carefully to ensure good adhesion between the repair material and the original surface. It is equally important to allow adequate time for a cleaned area to dry thoroughly before starting repair work. The best material for pipe repairs is "Sylmasta", a white ceramic plastic which can be bought from most hardware stores. It comes in two tins marked "A" and "B" and you have to mix equal parts from each tin and knead them together thoroughly before starting the repair job. If mending a stem you first roll out a piece of "Sylmasta" approximately the same length and thickness of the stem. This is best done by extruding the "Sylmasta" through a small mould such as a length of plastic cut from a ballpoint pen top. A *straight* piece of wire is then pushed through the extruded "sausage" and into the hole in the stem. By carefully working with dampened fingers a perfect joint can be obtained at the point of repair. Don't worry too much at this stage about shaping the stem. Wait until the "Sylmasta" has partially dried out (4 - 5 hours in a warm room) then carefully withdraw the wire. After 24 hours the "Sylmasta" will be quite hard and you can then sandpaper it to the finished shape required. If, on drying, the repair does not match the colour of the original pipe use white and grey watercolours to achieve a perfect blend.

To repair bowls you can either fix a large blob of "Sylmasta" over the damaged area and work on the fine detail with sandpaper, razor blades, and an assortment of scrapers when it is fully tried out, or you can, when repairing pipes with symetrical decorations, take a "Plasticine" impression from the opposite side and mould a piece of "Sylmasta" to fit the missing area. After final shaping this can be glued into shape using an epoxy-resin glue.

6 Sites for Pipe Hunting

So far as sites on which to hunt late Victorian pipes are concerned I can do no better than recommend to readers my earlier book, "Where to dig up Antiques", (Southern Collectors Publications) which provides a complete and *guaranteed* method for tracking down turn-of-the-century refuse dumps. The text of the book contains *everything* I have learned about site hunting during all my years of interest in the hobby; any further information I might attempt to add to it here would be mere repetition.

On the subject of sites for earlier clay pipes (16th century to early 19th century) my first recommendation to readers who would like to add some of these specimens to their collections is to become a member of your local archaeological society. By doing so you will gain ample opportunities to learn much more about locally-made pipes because every local society has at least one member who is an expert on the subject. He or she will be pleased to pass on knowledge and also to exchange duplicated specimens for the later pipes more likely to be found by those readers who are also bottle collectors.

On many post-medieval sites excavations often produce an abundance of pipes and those in charge of the work do not usually object to diggers keeping *a few* of their clay pipe finds - *after* the finds have been recorded. You can, of course, work on your own. If you live in a town or city with a history going back to medieval times your first step ought to be a visit to the local museum to study the pipes on display there. Note, in addition to the makers' marks and styles of decoration, the *locations* on which the pipes were found. Many will have turned up during demolition and site excavations within the town's boundaries and if other work takes place near these sites in the future it is highly likely that more pipes will be found. You can check the town's planning register to find out about proposed building and civil engineering projects and visit the locations when work commences. Examine the soil dug from trench walls and you will find your first early pipe very quickly. If the area is yielding large numbers of specimens offer to buy them from the workmen at a few pence each. Leave your name, address, and telephone number (on a business card if possible) and stress as you hand out this information that you are always interested in buying found objects. You might be pleasantly surprised by what is offered to you!

Anyone fortunate enough to live in a town or city with a river flowing through its centre will have no difficulty in finding early pipes, especially if the river is tidal. No matter how "industrialized" the present-day riverbanks are their tidal areas will produce excellent finds from all ages. You need only your eyes and a certain amount of patience to find clay tobacco pipe bowls on a tidal river. (See my earlier book, "Treasure Hunting For All". Blandford Press 1973. The text gives complete instructions for "eyes only" techniques). The Thames is by far the best river for pipe hunting; specimens can be found on both banks almost anywhere from Teddington Lock to Tilbury, the best locations being around the bridges between Kew and London Bridge. PLEASE do not be put off searching these foreshores by the thought that they must already have been searched by ten thousand pairs of eyes. It's true, they have: but EVERY tide brings new objects to light and the commonest objects (apart from rusty nails) are undoubtedly clay tobacco pipe bowls. Regular visits, especially those made on the lowest tides, cannot fail to produce early pipes for anyone who keeps head down and eyes open.

The banks of the Severn, the Dee, the Mersey, the Yorkshire Ouse, the Clyde, and the Forth are the *major* tidal riverside sites, but I stress that *every* river will produce finds for the diligent searcher. So too will canals for those prepared to work with rakes and (water conditions permitting) glass-bottomed buckets. So, too, will the stretch of water downstream from a bridge on many inland rivers far from built-up areas. If there is a long-established pub nearby then revellers in the past must have thrown pipes into the water ... GOOD HUNTING!

7 Valuing Pipes

I am extremely reluctant nowadays to quote "pounds and pence values" for any collectors' items, not only because of the problems galloping inflation presents, but also because sudden and often quite unexpected changes in collecting trends can make prices quoted one month look out-of-touch a month later. I have in the past tried to overcome these problems by using such ploys as an A-F rarity scale and a 1-200 price guide, both of which have failed to satisfy *all* who have consulted them, and I do not intend to compound my mistakes by offering a similar "solution" to the problem to present readers. Instead I offer a simple list which shows the comparative scarcity of the pipes mentioned throughout this book, based on *my* experience in hunting clay pipes and on *my* observations of pipes in other people's collections. At the top end of the list are the commonest, most-often found specimens. At the lower end are the extremely scarce specimens that have so far been found by only a handful of fortunate souls. Readers who seek up-to-the-minute information on prices are advised to consult dealers' lists, to attend auctions, and to scan the pages of all relevant magazines regularly. Only by doing all of these things can one expect to spot a bargain when it comes up!

The List

Plain bowls in late 19th and early 20th century specimens. (Those with broken stems are too common to merit inclusion even at the top of this list).
Basket bowls
R.A.O.B. bowls
Abstract patterned bowls
Plain bowls of 18th-mid 19th century types with *broken* stems.
Plain bowls with footballs beneath
Bowls with wheatsheaf designs back and/or front.
Bowls in the shape of trees; stumps
Bowls bearing regimental badges
Acorn bowls
Small claws holding bowls
Town and city coats-of-arms
Bowls decorated with fruit, flowers, or leaves
Bowls decorated with flags
Masonics - other than R.A.O.B.
Pre-1700 specimens with broken stems (short)
Pipes with complete stems bearing full names of makers (later than 1800)
Any of above with complete stems
Bowls with relief portraits of generals
Bowls with relief portraits of royalty, music hall stars, etc.
Figural bowls of Queen Victoria
Figural bowls of the Negro Boy
Pipes (including Dutch specimens) with complete decorated bowls and stems
Figurals advertising public houses
Skulls
Animals, including very large Claws
Specimens bearing dates, slogans, complete sentences
Stub-stemmed specimens of ornate design
Extra large specimens
Miniatures
French figurals
Pornographic specimens

8 Pipe Collections

Special outstanding collections of late 19th century clay tobacco pipes are owned by bottle collectors' club members-including June Heath, who kindly allowed us to photograph part of her collection for inclusion in this book. Consult current bottle collectors' literature and club newsletters to find other enthusiasts, most of whom will be delighted to show their collections, to swap, buy, and sell, and to encourage newcomers to share their interest.

Almost every museum in the land has *some* clay tobacco pipes either in the display cabinets or tucked away in the cellars. Drop a line to the curator of your local museum and you might find that an extensive collection of pipes from your area can be inspected within a few miles of your home. Outstanding collections including pipe makers' moulds and tools can also be seen at the following city, town, and county museums:

LONDON: Guildhall Museum, British Museum, London Museum, Cuming Museum, Bethnal Green Museum, Gunnersbury Park Museum.

BRISTOL: City Museum, Wills Museum.

Aberdeen	Edinburgh (National)	Peterborough
Alton	Grantham	Plymouth
Belfast	Leeds	Taunton
Birmingham	Leicester	Warrington
Boston	Lincoln	Winchester
Dartford	Nottingham	York (Castle Museum)

* Many tobacco companies also have extensive collections.

The Bottle Shop

80 NORTHAM ROAD, SOUTHAMPTON

TELEPHONE (0703) 23255

EUROPE'S LARGEST DISPLAY OF ANTIQUE BOTTLES, POT LIDS, DOLLS HEADS, PIPES, ETC.

BUY, SELL OR EXCHANGE

OPEN 6 DAYS PER WEEK — 9.30 A.M. TO 4.30 P.M.